HARCOURT

Math

Indiana ISTEP+ Test Prep

Grade 4

Harcourt

Orlando   Austin   Chicago   New York   Toronto   London   San Diego

Visit *The Learning Site!*
www.harcourtschool.com

CONTENTS

VOCABULARY REVIEW

Number Sense Vocabulary

Write the word or words that best complete
each sentence.

circle graph	digits	equivalent	hundredths	outcomes

1. A(n) _____ shows data as parts of a whole circle.

2. Two _____ are equally likely if they have the same chance of happening.

3. $\frac{2}{4}$ and $\frac{4}{8}$ are _____ fractions.

4. You read the decimal 0.09 as nine _____.

5. The symbols 0, 1, 2, 3, 4, 5, 6, 7, 8, 9, which make up numbers,

   are _____.

Match each word to its opposite.

6. **unlikely** •   • **denominator**

7. **numerator** •   • **exact**

8. **even** •   • **likely**

9. **rounded** •   • **impossible**

10. **certain** •   • **odd**

Name _____

Computation Vocabulary
Circle the word or words that answer the question.

1. What do you do when you combine equal groups?

   multiply   **subtract**

2. How can you check your answer to an addition problem?

   divide   **estimate**

3. Which term describes the relationship between multiplication and division?

   remainder   **inverse operations**

4. What do you do when you separate objects into equal groups?

   divide   **add**

5. What is a set of related multiplication and division sentences called?

   fact family   **multiples**

6. What is it called when you change 14 ones to 1 ten 4 ones?

   regrouping   **rounding**

7. What property states that any number times 1 equals that number?

   Identity Property   **Zero Property**

8. What does 4 represent in the equation 24 ÷ 6 = 4?

   factor   **quotient**

9. What is the equation 7 × 0 = 0 an example of?

   Zero Property   **Distributive Property**

10. What do you call a symbol or letter that stands for an unknown number?

   divisor   **variable**

Name _____

Algebra and Functions Vocabulary
Choose the word or words that better complete each sentence. Then write the word or words on the line.

1. $12 + 6 = 18$ is an example of a(n) _____.

   equation　　**fact family**

2. An _____ is an arrangement of objects in rows and columns.

   addend　　**array**

3. 2, 20, 200 is an example of a _____.

   pattern　　**schedule**

4. $7 + 5$ is a(n) _____.

   expression　　**number sentence**

5. 10, 20, 30, and 40 are all _____ of 10.

   factors　　**multiples**

Choose the property that matches each description.

Associative	**Commutative**	**Distributive**	**Identity**	**Zero**

6. You can think of one factor as the sum of two addends. For example, $3 \times 6 = (3 \times 2) + (3 \times 4)$.

   _____ Property

7. The product of 1 and any number equals that number. For example, $1 \times 7 = 7$.

   _____ Property

8. You can group factors in different ways and get the same answer. For example, $(2 \times 4) \times 3 = 2 \times (4 \times 3)$.

   _____ Property

9. You can multiply two factors in any order and get the same product. For example, $3 \times 4 = 4 \times 3$.

   _____ Property

10. The product of 0 and any number equals 0. For example, $3 \times 0 = 0$.

    _____ Property

Geometry Vocabulary

Write the word or words that best complete each sentence.

congruent	line	line of symmetry	obtuse angle	quadrilateral

1. Figures that have the same size and shape are _____.

2. A(n) _____ has a measure greater than a right angle.

3. A(n) _____ continues in both directions and does not end.

4. A shape can be divided into two matching parts by drawing a(n) _____.

5. A polygon with four sides is called a(n) _____.

Name _____

Geometry Vocabulary
Match the clue to the shape.

1. I have one **face** that is a **circle**.   •

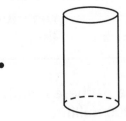

•

2. I have two **faces**, no **edges,** and no **vertices**.   •

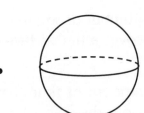

•

3. I have 6 **faces** that are all **squares**.   •

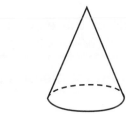

•

4. I have no **faces**, no **edges,** and no **vertices**.   •

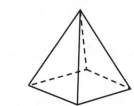

•

5. I have 8 **edges** and 5 **vertices**.   •

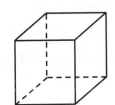

•

Name _____

Measurement Vocabulary

Write the word or words that match each definition.

| area | capacity | degrees Fahrenheit | elapsed time | gallon |
| kilogram | liter | perimeter | quart | volume |

1. one of these equals
   1,000 milliliters   _____

2. number of squares that
   cover a flat surface   _____

3. amount of time that passes
   from the start of an activity
   to the end of that activity   _____

4. one of these equals two pints   _____

5. a unit for measuring temperature   _____

6. amount of space a solid figure
   takes up   _____

7. one of these equals four quarts   _____

8. metric unit for measuring mass   _____

9. distance around a figure   _____

10. amount a container can hold   _____

Getting Ready for the **ISTEP**+

1 Indiana has 6<u>6</u>3 towns, villages, and cities. What is the value of the underlined digit 6?

(A) 6

(B) 60

(C) 600

(D) 6,000

2 Indiana has 89 museums. Jane's family wants to visit all of them. So far they have visited 27 museums. Which number sentence could you use to find the number of museums they have left to visit?

(A) 89 + 27 = 116

(B) 89 − 27 = 62

(C) 116 + 27 = 143

(D) 116 − 89 = 27

GO ON ▶

3 The table below shows the areas of three Indiana counties.

INDIANA COUNTIES	
County	Area (in square miles)
Jackson	514
Porter	522
Washington	517

Which shows the areas in square miles in order from LEAST to GREATEST?

Ⓐ 514, 517, 522

Ⓑ 522, 517, 514

Ⓒ 517, 514, 522

Ⓓ 514, 522, 517

GO ON ▶

© Harcourt

4 Ted needs 60 pounds of grapes for a club picnic. Use rounding to ESTIMATE the sum of each pair of grape cartons. Then decide which pair he should buy to have the least amount left over.

(A)

17 pounds   49 pounds

(B)

34 pounds   17 pounds

(C)

47 pounds   28 pounds

(D)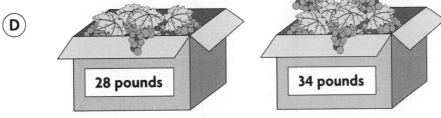

28 pounds   34 pounds

© Harcourt

GO ON ▶

5 Erica is driving from New Albany to Michigan City. The distance is 262 miles. She stops for a break after driving 138 miles. How many more miles does she have to drive?

Write a number sentence that shows how many more miles she has to drive.

_____

Use your number sentence to find how many more miles she has to drive.

Show All Work

Answer _____ more miles

6 How many feet higher is Churchill Peaks, Alaska, than Hoosier Hill, Indiana?

STATES' HIGHEST POINTS	
Location	Highest Point (in feet)
Churchill Peaks, Alaska	20,320
Hoosier Hill, Indiana	1,257
Mount Arvon, Michigan	1,979
Mount Elbert, Colorado	14,433

A 13,176 feet

B 18,341 feet

C 19,063 feet

D 21,577 feet

GO ON ▶

7 Harris has the bills and coins shown below.

He wants to buy 2 markers that cost $1.79 each.
Does he have enough money?

Show All Work

_____

_____

_____

Draw the coins Harris will have left after he pays for
the markers.

How much money will Harris have left?

Answer _____

GO ON ▶

8 Which clock shows 7:35?

Ⓐ

Ⓑ

Ⓒ

Ⓓ

9 Which number sentence matches?

$4 + 4 + 4 + 4 + 4$

Ⓐ $5 \times 5 = 25$

Ⓑ $3 \times 4 = 12$

Ⓒ $4 \times 4 = 16$

Ⓓ $5 \times 4 = 20$

GO ON ▶

10 Which number sentence does the number line show?

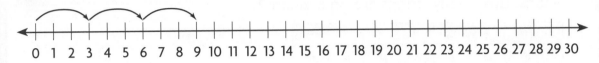

0 1 2 3 4 5 6 7 8 9 10 11 12 13 14 15 16 17 18 19 20 21 22 23 24 25 26 27 28 29 30

(A) $3 + 9 = 12$

(B) $3 \times 1 = 3$

(C) $3 + 3 = 6$

(D) $3 \times 3 = 9$

11 A community center is planning a camping trip to Shakamak State Park. The available tents sleep 5 people each. The community center owns 7 tents. How many people can sleep in the tents on the camping trip?

(A) 12 people

(B) 35 people

(C) 53 people

(D) 75 people

GO ON ▶

12 The table shows how many strings there are on violins. How many strings are on 6 violins?

Violins	1	2	3	4	5	6
Strings	4	8	12	16	20	■

Ⓐ 28 strings

Ⓑ 26 strings

Ⓒ 24 strings

Ⓓ 22 strings

13 Which number is missing in the number sentence?

$4 \times 9 = ■ \times 4$

Ⓐ 4

Ⓑ 9

Ⓒ 13

Ⓓ 36

© Harcourt

GO ON ▶

14 Movie tickets cost $6.50 each. Doreen and her younger sister Sheila go to the movies. They pay with a $20.00 bill. Explain how you would find the amount of change they will get.

_____

_____

_____

How much change will they get?

Answer _____

---

15 Which number sentence is related to 20 ÷ 4 = 5?

(A) $4 + 5 = 9$

(B) $5 \times 4 = 20$

(C) $20 - 5 = 15$

(D) $20 \times 4 = 80$

GO ON ▶

16 There are 24 students going to the Joseph Moore Museum in Richmond, Indiana. Each van can hold 6 students. Write a division sentence to find the number of vans that are needed.

_____

Solve your division sentence to find the number of vans that are needed.

Answer _____ vans

17 Which symbol is needed to make the number sentence true?

32 ● 4 = 8

(A) +

(B) −

(C) ×

(D) ÷

GO ON ▶

18 Use a ruler to measure the length to the nearest half inch.

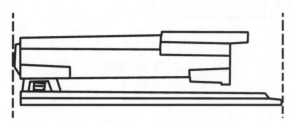

(A) 2 inches

(B) $2\frac{1}{2}$ inches

(C) 3 inches

(D) $3\frac{1}{2}$ inches

19 Which figure shows a line of symmetry?

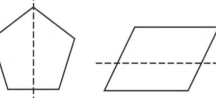 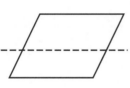 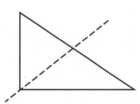 

(A)   (B)   (C)   (D)

GO ON ▶

20 Find the perimeter.

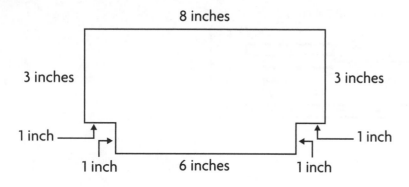

(A) 16 inches

(B) 20 inches

(C) 24 inches

(D) 28 inches

21 Which pattern can be described by this rule?

Rule: Multiply the number of squares in a row by 3.

Ⓐ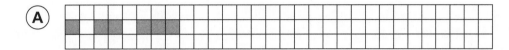

Ⓑ

Ⓒ

Ⓓ

GO ON ▶

22 Which fraction has a numerator of 4 and a denominator of 5?

(A) $\frac{4}{5}$

(B) $\frac{5}{4}$

(C) $\frac{4}{9}$

(D) $\frac{5}{9}$

23 Paul ate $\frac{1}{8}$ of a pizza. Greg ate $\frac{4}{8}$ of the pizza. How much of the pizza is left?

(A) $\frac{3}{8}$

(B) $\frac{5}{8}$

(C) $\frac{6}{8}$

(D) $\frac{7}{8}$

GO ON ▶

24 Which fraction and decimal does the model show?

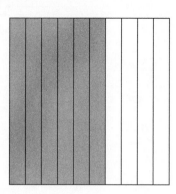

(A) $\frac{9}{10}$, 0.9

(B) $\frac{6}{10}$, 0.6

(C) $\frac{3}{10}$, 0.3

(D) $\frac{1}{10}$, 0.1

GO ON ►

25 Draw a diagram to show how to use counters to find the quotient 17 ÷ 3.

Now write the quotient and the remainder.

Answer 17 ÷ 3 = _____

STOP ■

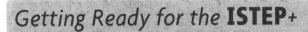 *Getting Ready for the* **ISTEP+**

---

1 What is the word form for 330,030?

(**A**) three hundred thousand, thirty

(**B**) three hundred thirty thousand

(**C**) three hundred thirty thousand, thirty

(**D**) three hundred three thousand, thirty

---

2 On the chalkboard, Jesse writes the following in standard form.

200,000 + 5,000 + 40 + 9

What number does Jesse write?

(**A**) 200,549

(**B**) 205,049

(**C**) 205,490

(**D**) 250,409

GO ON ▶

3 The pictograph below shows the average life spans of some animals.

Average Life Spans of Some Animals	
Animal	**Number of Years**
Baboon	▦ ▦ ▦ ▦
Grizzly Bear	▦ ▦ ▦ ▦ ▦
Cow	▦ ▦ ▦
Wolf	▦
African Elephant	▦ ▦ ▦ ▦ ▦ ▦ ▦

Key: Each ▦ = 5 years.

On the lines below, tell how you can find out how much longer a grizzly bear lives than a cow.

_____

_____

_____

_____

_____

_____

_____

_____

The average life span of an Asian elephant is about 25 years longer than the life span of a cow.

How many ▦ would you need to show the life span of an Asian elephant on this graph?

Answer _____ ▦

GO ON ▶

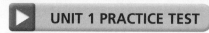

4 For which ⬤ can you write <?

(A) 32,145 ⬤ 32,145

(B) 978,998 ⬤ 978,989

(C) 345,076 ⬤ 345,067

(D) 857,970 ⬤ 875,087

5 Which number on the number line is LESS THAN 6,720?

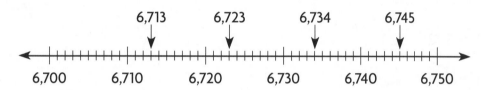

(A) 6,713

(B) 6,723

(C) 6,734

(D) 6,745

GO ON ▶

6 The table below shows the number of passengers on the morning train from Monday through Wednesday.

TRAIN PASSENGERS	
Day	Passengers
Monday	342
Tuesday	324
Wednesday	350

Order the number of passengers who ride the train each morning from GREATEST to LEAST.

(A) 342 < 324 < 350

(B) 324 > 342 > 350

(C) 350 > 342 > 324

(D) 350 < 324 < 342

GO ON ▶

7 Endangered animals in the United States include 65 species of mammals, 78 species of birds, 71 species of fish, and 35 species of insects. Which table shows this data organized from LEAST to GREATEST?

A

ENDANGERED ANIMALS	
Species	**Number**
Insects	35
Mammals	65
Fish	71
Birds	78

C

ENDANGERED ANIMALS	
Species	**Number**
Birds	78
Mammals	65
Fish	71
Insects	35

B

ENDANGERED ANIMALS	
Species	**Number**
Birds	78
Fish	71
Mammals	65
Insects	35

D

ENDANGERED ANIMALS	
Species	**Number**
Mammals	65
Fish	71
Birds	78
Insects	35

GO ON ▶

8 The five tallest mountains in the United States, Canada, and Mexico in order of height from the tallest to the shortest are McKinley, Logan, Pico de Orizaba, St. Elias, and Popocatepetl. The heights of the mountains in random order are 19,551 feet, 18,008 feet, 20,320 feet, 18,555 feet, and 17,930 feet.

Make a table to show the mountains and their heights in order from TALLEST to SHORTEST.

On the lines below, explain how someone would know from looking at the table that Logan is taller than Pico de Orizaba.

_____

_____

_____

GO ON ▶

9 Mr. Marshall is a pilot. Last month he flew a total distance of 8,900 miles, which is rounded to the nearest hundred miles. How many miles could he have flown last month?

(A) 7,999 miles

(B) 8,829 miles

(C) 8,874 miles

(D) 8,969 miles

10 Use mental math to find the difference between 900 and 500.

(A) 1,400

(B) 1,200

(C) 600

(D) 400

GO ON ▶

11 Winston Elementary School has 802 students. King Elementary School has 689 students. ABOUT how many more students does Winston have than King?

(A) about 100 more students

(B) about 400 more students

(C) about 1,400 more students

(D) about 1,500 more students

12 The map below shows some driving distances.

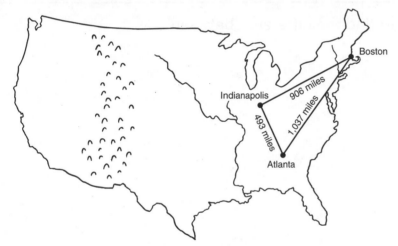

The Perez family traveled a total of 1,530 miles. Which trip could they have taken?

(A) from Boston to Indianapolis and from Indianapolis to Atlanta

(C) from Atlanta to Indianapolis and from Indianapolis back to Atlanta

(B) from Boston to Atlanta and from Atlanta to Indianapolis

(D) from Boston to Atlanta and from Atlanta back to Boston

GO ON ▶

13 Mrs. Wallace and her children are driving 3,005 miles to California. So far they have driven 1,965 miles. How many miles away from California are they?

(A) 1,040 miles

(B) 2,960 miles

(C) 3,965 miles

(D) 4,970 miles

14 Darryl scored 1,005 points on a video game. His friend Hal scored 94 fewer points. Which is Hal's score?

(A) 119

(B) 911

(C) 1,099

(D) 2,094

© Harcourt

GO ON ▶

15 Which of the following shows a reasonable method and correct solution for 763,078 + 139,987?

(A) method: mental math;
solution: 905,065

(B) method: pencil and paper;
solution: 905,065

(C) method: mental math;
solution: 903,065

(D) method: pencil and paper;
solution: 903,065

16 On Monday 987 tickets were sold for the senior-class play to be performed in the school auditorium. The auditorium has 1,015 seats. Do you need an estimate or an exact answer to tell how many more tickets to the play can be sold? Explain.

_____

_____

_____

_____

How many more tickets can be sold to the play?

Answer _____ more tickets

GO ON ▶

17 Pablo's father gave him 36 trading cards. Then Pablo bought more cards at a card show. He sold 6 of the new cards he bought. If n represents the new cards he bought, which expression shows the number of cards Pablo has now?

(A) $(36 - n) + 6$

(B) $36 - (n + 6)$

(C) $36 + (n - 6)$

(D) $36 + (n + 6)$

18 Sara bought a bag of 18 apples. The bag tore and 6 apples fell out. Which equation can you use to find the number of apples left in the bag? Use x for the number of apples left in the bag.

(A) $x - 6 = 18$

(B) $18 - 6 = x$

(C) $18 + x = 6$

(D) $18 + 6 = x$

GO ON ▶

19 The rule is add 9. The equation is $n + 9 = m$.
Which two numbers are missing in the table?

INPUT	n	18	26	15	37	43
OUTPUT	m	27	35	24	■	■

(A) 28, 34

(B) 37, 43

(C) 46, 52

(D) 47, 53

20 Gina has 15 pennies in her bank. She puts in
5 more pennies and then takes out 7 pennies.
Her brother Jeff has 6 pennies in his bank. How
many more pennies does he need to have the
same number that Gina has? Explain what steps
you could use to solve the problem. Then find
the answer.

_____

_____

_____

_____

_____

Now find the answer.

Answer Jeff needs _____ more pennies.

© Harcourt

STOP ■

Name _____

Getting Ready for the ISTEP+

---

1 The school bus picks up Janine each morning at 7:52 A.M. Her bus ride takes 25 minutes. What time does Janine arrive at school?

(A) 8:00 A.M.

(B) 8:07 A.M.

(C) 8:17 A.M.

(D) 8:25 A.M.

---

2 Howard takes a bus from Evansville to Bloomington. The bus leaves Evansville at 3:49 P.M. The ride takes 2 hours and 35 minutes. What time will Howard arrive in Bloomington?

(A) 5:35 P.M.

(B) 6:24 P.M.

(C) 6:35 P.M.

(D) 7:01 P.M.

GO ON ▶

For 3–4, use the schedule of activities at Hoosier National Forest.

Hoosier National Forest Activities

Activity		Time
Fishing	(Open)	6:00 A.M. to 9:00 P.M.
Watchable Wildlife Tours	(2 Hours)	8:00 A.M.; 7:00 P.M.
Mountain Biking	(3 Hours)	9:00 A.M.; 10:00 A.M.; 11:00 A.M.; 2:00 P.M.; 3:00 P.M.
Hike	(2 Hours)	9:30 A.M.; 1:00 P.M.
Horseback Riding	(50 minutes)	1:30 P.M.; 2:30 P.M.; 3:30 P.M.

3 Hong wants to go horseback riding, fishing, and take the Watchable Wildlife Tour today. Her family is arriving at the park at 8:00 A.M. and having a picnic at noon. They are leaving the park at 3:00 P.M. Complete Hong's schedule to show how she can attend all three activities.

HONG'S SCHEDULE	
Time	**Activity**
12 noon	Picnic
3:00 P.M.	Leave

4 If Hong wants to go mountain biking, which activities would she have to give up?

(A) Watchable Wildlife Tour and fishing

(B) Watchable Wildlife Tour and horseback riding

(C) Fishing and horseback riding

(D) Watchable Wildlife Tour and picnic

GO ON ▶

5 Mr. Pinter's students planted peony bushes in the fall. In the spring, the students counted the number of peonies that bloomed on each bush. The tally table shows their results.

How many peonies bloomed in all?

PEONY BLOOMS	
Bush	**Number Counted**
Bush 1	IIII
Bush 2	ЖЖ II
Bush 3	ЖЖ I
Bush 4	ЖЖ ЖЖ II

(A) 25 peonies

(B) 27 peonies

(C) 29 peonies

(D) 31 peonies

6 The line plot shows the heights of cornstalks Damien grew in his backyard garden.

How many cornstalks grew to be at least 6 feet tall?

(A) 3 cornstalks

(B) 5 cornstalks

(C) 8 cornstalks

(D) 10 cornstalks

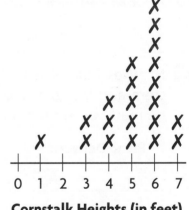

Cornstalk Heights (in feet)

© Harcourt

GO ON ▶

7 Mara's line plot shows the number of students absent from her class for one week.

Which BEST describes the pattern you see in the data?

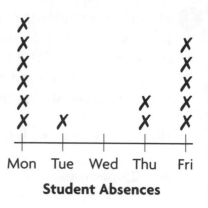

Student Absences

(A) Fewer students were absent at the beginning of the week than at the end of the week.

(B) Fewer students were absent at the end of the week than in the middle of the week.

(C) More students were absent on Mondays and Fridays than on the other days.

(D) More students were absent during the middle of the week than at the beginning and end of the week.

---

8 Mr. Westfall delivers pianos for a music store. The stem-and-leaf plot shows the number of deliveries he made each month last year.

Piano Deliveries

Stem	Leaves
1	3 7
2	0 5 8
3	0 2 6 9
4	0 1 2

1|7 = 17 deliveries.

What is the median number of deliveries he made?

30	31	36	39
(A)	(B)	(C)	(D)

GO ON ▶

© Harcourt

9 Barton Elementary School had a recycling drive. The bar graph shows the number of cans each grade collected.

What is the interval of the graph?

(A) 100 (C) 400

(B) 200 (D) 600

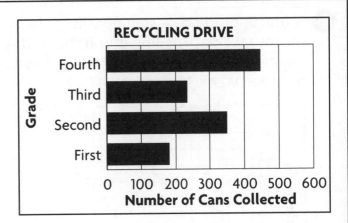

RECYCLING DRIVE

Grade — Fourth, Third, Second, First

Number of Cans Collected: 0 100 200 300 400 500 600

10 There were many different sizes of dinosaurs. Stegosaurus was about 30 feet long. Tyrannosaurus was about 40 feet long. Gallimimus, or "ostrich dinosaur," was about 20 feet long. Brachiosaurus was about 80 feet long. Triceratops was about 30 feet long.

Use the data to complete the pictograph.

DINOSAUR LENGTHS

Type of Dinosaur	Length (in feet)
Brachiosaurus	
Gallimimus	
Stegosaurus	
Triceratops	
Tyrannosaurus	

Key: Each ▽ = 10 feet.

Which two dinosaurs have the greatest difference in length? Use your pictograph to answer the question.

Answer _____

GO ON ▶

11 A fourth-grade class sold pencils in the school yard for four weeks. The line graph shows the total sales each week. What sales trend do you see between week 1 and week 4?

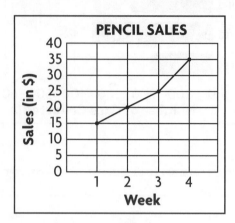

A Sales went up.

B Sales went down.

C Sales stayed the same.

D Sales went down at first, and then went up.

12 The line graph shows attendance at the school play at Lincoln Elementary School. On which day was attendance the lowest?

A Thursday

B Friday

C Saturday

D Sunday

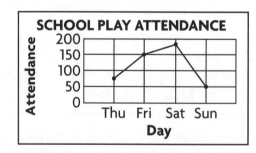

GO ON ▶

© Harcourt

13 Marsha ran a 4-mile race in 32 minutes. The line graph shows her time.

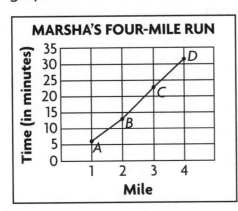

Which point on the graph shows that Marsha had run 3 miles in 23 minutes?

A point *A*

B point *B*

C point *C*

D point *D*

GO ON ▶

14 An African elephant needs up to 24 gallons of water a day. The table below shows how much water Daisy, an African elephant, drank over five days.

WATER DAISY DRANK	
Day	**Amount (in gallons)**
Monday	24
Tuesday	19
Wednesday	13
Thursday	20
Friday	22

Use the information in the table to make a line graph.

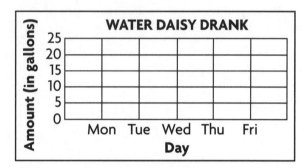

Use your graph to answer the questions below.

Was the total amount Daisy drank over five days GREATER THAN or LESS THAN 100 gallons?

Answer _____

On which day did Daisy drink the least amount of water?

Answer _____

GO ON ▶

15 Ronnie asked his friends to name their favorite fruits. He recorded the answers in a table. Then he made this circle graph from his data.

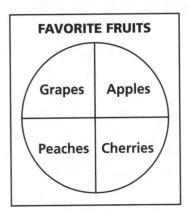

FAVORITE FRUITS

Grapes | Apples

Peaches | Cherries

Which of these is Ronnie's table?

Ⓐ
FAVORITE FRUITS	
Fruit	Votes
Apples	2
Grapes	4
Peaches	6
Cherries	8

Ⓒ
FAVORITE FRUITS	
Fruit	Votes
Apples	4
Grapes	5
Peaches	8
Cherries	6

Ⓑ
FAVORITE FRUITS	
Fruit	Votes
Apples	8
Grapes	8
Peaches	8
Cherries	8

Ⓓ
FAVORITE FRUITS	
Fruit	Votes
Apples	8
Grapes	4
Peaches	4
Cherries	8

GO ON ▶

16 The table below shows how much Steve's parents spent for his birthday party.

PARTY COSTS	
Item	Cost
Pizza	$40
Cake	$15
Soda	$20
Decorations	$15
Door Prizes	$10

If you make a circle graph using the data, how many sections will the circle graph have?

(A) 2 sections

(B) 3 sections

(C) 4 sections

(D) 5 sections

17 Tim's parents measure his height every year. What kind of graph would best display this data?

(A) line plot

(B) stem-and-leaf plot

(C) circle graph

(D) line graph

© Harcourt

GO ON ▶

18 Elaine plays on the youth basketball team. She made 7 baskets in the first game. She did not make any baskets in the second game. She made 10 baskets in the third game. She made 12 baskets in the fourth game.

Which graph most clearly shows the data?

A

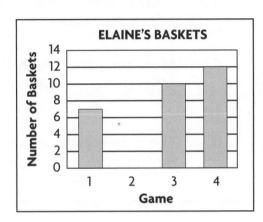

B

C

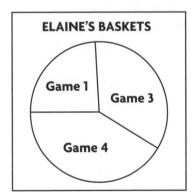

D

Elaine's Baskets	
Stem	**Leaves**
0	0 7
1	0 2

1|2 = 12 baskets.

GO ON ▶

19 The line graph shows about how many swimsuits a clothing store sells during each season.

Which conclusion would you NOT draw from the graph?

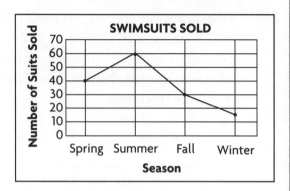

SWIMSUITS SOLD

Number of Suits Sold

Season

A Spring is the most popular time to buy a swimsuit.

C More suits are sold in summer than in winter.

B More than 50 suits are sold during the summer months.

D The fewest suits are sold during the winter months.

20 A new community Sports Center has a pool, a track, an exercise room, and tennis courts. The bar graph shows the number of people who used the exercise room and who played tennis on opening day. In addition, 60 people went swimming and 35 people ran around the track. Use the data to complete the graph.

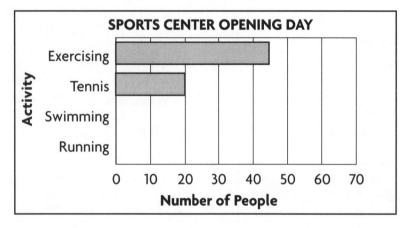

SPORTS CENTER OPENING DAY

Activity

Exercising
Tennis
Swimming
Running

Number of People

STOP ■

© Harcourt

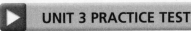

Getting Ready for the **ISTEP+**

---

1 Ms. Kelly's fourth-grade class has 5 rows of desks. There are 4 desks in each row. How many desks are in Ms. Kelly's class?

(**A**) 9 desks

(**B**) 15 desks

(**C**) 20 desks

(**D**) 25 desks

---

2 Which is a related equation for $12 \div 2 = 6$?

(**A**) $2 \times 12 = 24$

(**B**) $1 \times 12 = 12$

(**C**) $4 \times 3 = 12$

(**D**) $2 \times 6 = 12$

GO ON ▶

3 Mrs. Lawry has 24 students in her class. There are 4 reading groups. Each group has the same number of students. How many students are in each reading group?

(A) 4 students

(B) 5 students

(C) 6 students

(D) 8 students

4 Which pair of related equations is represented by the model below?

(A) $3 + 4 = 7$ and $7 - 4 = 3$

(B) $4 \times 4 = 16$ and $16 \div 4 = 4$

(C) $3 \times 4 = 12$ and $16 \div 4 = 4$

(D) $3 \times 4 = 12$ and $12 \div 3 = 4$

GO ON ▶

5 Which pair of arrays could you use to find the product 4×10?

10
4

A
5   5
4   4

B
10   10
3   3

C
5   5
2   2

D
5   5
5   5

GO ON ▶

6 What is the eighth number in the pattern?

11, 22, 33, 44, 55, . . .

(A) 66   (B) 77   (C) 88   (D) 99

7 Which property would you use to find 1×17?

(A) Associative Property

(B) Zero Property

(C) Commutative Property

(D) Identity Property

8 The Sports Outlet has 5 bicycles on display. Each bicycle has 2 pairs of streamers. None of the bicycles has headlights. There are 4 reflectors on each bicycle. Which equation shows the total number of headlights on all the bicycles?

(A) $5 \times 0 = 0$

(B) $5 \times 1 = 5$

(C) $5 \times 2 = 10$

(D) $5 \times 5 = 25$

GO ON ▶

9 Lori is buying juice boxes for her party. She is having 28 guests.

JUICE BOXES		
Package Size	Boxes per Package	Cost per Package
Medium	6	$2

If Lori buys 4 medium packages of juice boxes, will she have enough so that each guest will have a box of juice?

On the lines below, tell which operation is needed to solve the problem. Explain why. Then solve.

_____

_____

_____

_____

Answer

_____

Lori decides to spend $10 on juice boxes. How many packages does she buy? How many juice boxes does she buy?

Show your work.

Answer _____ packages

_____ juice boxes

GO ON ▶

10 A toy store has these items in the window.

Mr. Rios bought each of his grandchildren a set of watercolors. He spent $24 in all. How many grandchildren does Mr. Rios have?

On the lines below, tell which operation you can use to solve this problem and explain why.

_____

_____

_____

_____

Write a number sentence and use it to solve the problem.

Answer

_____

GO ON ▶

© Harcourt

11 The map shows the distances between four places.

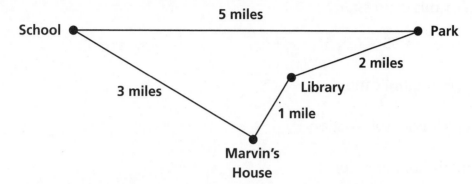

Marvin rode his bike from his house to school and then back home each of the 5 days of the school week. On Friday, he also rode from his house to the library and back home. How many miles did he ride in all?

Write an expression to match the words. Then find the value of the expression.

Answer

_____

_____

Sophie rode from school to the park and then back to school. Then she rode to Marvin's house. How many miles did she ride in all?

Write an expression to match the words. Find the value of the expression.

Answer

_____

_____

GO ON ▶

12 Which is the correct order of operations to find the value of this expression?

$$35 - 6 \times 5$$

(A) division, multiplication

(B) multiplication, subtraction

(C) multiplication, addition

(D) subtraction, multiplication

13 Use the order of operations to find the value of this expression.

$$56 \div (4 + 3)$$

(A) 8

(B) 12

(C) 14

(D) 17

GO ON ▶

14 In 2000, there were about 64,000 farms in Indiana. By 2001, there were a fewer number of farms, f. Which expression can you use to find how many farms there were in 2001?

(A) $64,000 + f$

(B) $64,000 - f$

(C) $f - 64,000$

(D) $f \div 64,000$

15 Melissa divided some books among the 5 children at the book fair. Each child received 4 books. Which equation could you use to find the total number of books?

(A) $4 + k = 5$

(B) $k \div 5 = 4$

(C) $4 \times k = 5$

(D) $k - 5 = 4$

GO ON ▶

16 Brenda saved the money she received for her
birthday from her grandparents. She also saved
$3 each week from her allowance. After 4 weeks,
she had saved a total of $37. How much did Brenda
receive as a birthday gift from her grandparents?

Write an equation. Then work backward to solve
the equation.

Show your work.

Answer _____

GO ON ▶

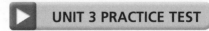
17 What happens to the values of the sides of the equation if you multiply both sides by 2?

$$(4 + 3) = (35 \div 5)$$

(A) The values of both sides decrease by 2.

(B) The values of both sides increase by 2.

(C) The values of both sides double.

(D) The value of both sides stay the same.

18 Which equation is the rule you can use to find the missing number?

INPUT	r	9	5	4	2
OUTPUT	s	■	35	28	14

(A) $r \times 7 = s$

(B) $s \times 7 = r$

(C) $r \div 7 = s$

(D) $s + 7 = r$

GO ON ▶

19 A rule for an input/output table is $a = 6 \times b$. What is the value of a when $b = 3$?

(A) $a = 2$

(B) $a = 3$

(C) $a = 9$

(D) $a = 18$

20 Use the equation $b = a - 4$ to find the missing number.

INPUT	a	8	10	12	14
OUTPUT	b	4	6	8	■

(A) 8

(B) 9

(C) 10

(D) 12

STOP ■

Getting Ready for the **ISTEP+**

1 What is the value of *n* in this equation?

$$9 \times n = 4{,}500$$

(A) *n* = 5

(B) *n* = 50

(C) *n* = 500

(D) *n* = 5,000

2 Each box holds 30 CDs.

Number of Boxes	1	2	3	4	5	6
CDs in Each Box	30	60	90	■	■	■

Which set of numbers completes the table?

(A) 91, 92, 93

(B) 30, 60, 90

(C) 120, 150, 180

(D) 400, 450, 500

GO ON ▶

3 Ms. White buys 9 pairs of jeans for her children. She pays $27.59 for each pair. Which is a reasonable estimate for the total cost of the jeans?

(A) about $270

(B) about $200

(C) about $180

(D) about $90

4 Find the product.

$$\begin{array}{r} 73 \\ \times\ 6 \\ \hline \end{array}$$

(A) 439

(B) 438

(C) 429

(D) 428

GO ON ▶

5 Tim and his dad have been to 7 Indianapolis Colts football games during the past season. They live 32 miles from the football stadium. Tim figured out that they traveled 448 miles in all. Is that REASONABLE?

(A) No. 7 × 32 is about 7 × 30, or 210 miles.

(B) No. Tim is just guessing. There is no way to know how far they traveled.

(C) No. The answer should be about 320 miles.

(D) Yes. A round trip is about 60 miles and 7 × 60 = 420 miles.

GO ON ▶

6 Tom is building 4 new bookcases for the library. He needs 48 screws for each bookcase. How many screws does he need for all 4 bookcases?

(A) 168 screws

(B) 192 screws

(C) 208 screws

(D) 218 screws

7 Mr. and Mrs. Anderson and their 3 children are taking a train trip to visit their relatives. Each ticket costs $80. What is the total cost of the tickets?

(A) $400

(B) $320

(C) $240

(D) $160

GO ON ▶

8 An 8-ounce container of strawberry yogurt has 230 calories. How many calories are there in 3 containers?

Write a number sentence, and then solve the problem.

Show All Work

Answer _____ calories

9 Use a basic fact to find the product.

8 × 30

(A)   24

(B)   240

(C)   2,400

(D) 24,000

GO ON ▶

10 Shade the grid below to find the product 7 × 26.

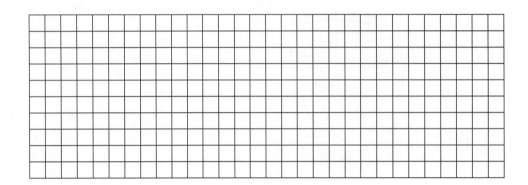

On the lines below, explain how to use a model to multiply 7 × 26.

_____

_____

_____

_____

Use the Distributive Property to show the multiplication as the sum of two products. Find the sum.

Show All Work

Answer _____

GO ON ▶

11 Find the product.

$$\begin{array}{r} 37 \\ \times\ 10 \\ \hline \end{array}$$

(A) 37

(B) 137

(C) 370

(D) 3,700

12 Last year, each student and teacher at Lakeview School used an average of 18 pounds of paper. There are a total of 314 students and teachers in the school. ESTIMATE how much paper the school used last year.

(A) about 300 pounds

(B) about 600 pounds

(C) about 3,000 pounds

(D) about 6,000 pounds

© Harcourt

GO ON ▶

13 Kate wrote the following solution to 70×346.

$$346 = 300 + 40 + 6$$

$$\underline{\times\ 70} \qquad\qquad 70 \leftarrow 70 \times (300 + 40 + 6)$$

$$21{,}000 \leftarrow 70 \times 300$$

$$2{,}800 \leftarrow 70 \times 40$$

$$\underline{+\ \ \ \ 420} \leftarrow 70 \times 6$$

$$24{,}220$$

On the lines below, explain whether or not this is a REASONABLE solution.

_____

_____

_____

_____

Use Kate's method to multiply 30×823.

Show Your Work

GO ON ▶

14 Find the value for *n* that makes the equation true.

$$74 \times n = 14{,}800$$

(A) $n = 175$

(B) $n = 200$

(C) $n = 225$

(D) $n = 250$

15 Which estimate would you use to check this expression?

$$324 \times 47$$

(A) $320 \times 40 = 12{,}800$

(B) $330 \times 40 = 13{,}200$

(C) $320 \times 50 = 16{,}000$

(D) $330 \times 50 = 16{,}500$

GO ON ▶

16 Which of the following describes the mistake in solving this multiplication problem?

$$
\begin{array}{r}
7{,}248 \\
\times\quad 43 \\
\hline
21\ 744 \\
28\ 992 \\
\hline
50{,}736
\end{array}
$$

(A) tens multiplied incorrectly

(B) incorrect regrouping when adding the products

(C) incorrect regrouping when multiplying by ones

(D) zero missing from second partial product

17 Which problem would you MOST LIKELY solve using mental math?

(A) $1{,}200 \times 40$

(B) $3{,}478 \times 61$

(C) $6{,}999 \times 82$

(D) $8{,}004 \times 77$

GO ON ▶

18 Bradley saved $12.50 each month for 18 months. Then he spent $189 to go to music camp. How much money does Bradley have left?

(A) $25

(B) $36

(C) $64

(D) $89

19 Julia owns a sandwich wagon. She sets up her business every day at noon in the park.

Julia's Lunches	
Hamburger	$4.75
Ham & Cheese	$5.99
Egg Salad	$3.49

Julia sold 22 hamburgers and 32 ham and cheese sandwiches on Tuesday. Write an equation you could use to find the amount of money Julia took in from her sales. Then solve the problem.

Show All Work

Answer _____

GO ON ▶

20 The Pickens family went to an amusement park. They bought 2 adult tickets, 1 child ticket, and 3 student tickets.

Amusement Park	
Ticket	Price
Adult	$11.95
Child (under 6)	$ 8.25
Student	$10.50

If they gave the cashier $75.00, how much money did they get back?

(A) $10.35

(B) $11.35

(C) $19.60

(D) $63.65

STOP ■

Getting Ready for the **ISTEP+**

1 There are 20 members of the Glendale Marching Band. The leader wants the band to form rows with 3 members in each row. How many rows will there be? How many members will be left over?

Draw a model using counters to find out how many equal groups of 3 are in 20 and how many are left over.

How many rows of 3 can the band members form? How many band members are left over?

Answer _____ rows with _____ band members left over

GO ON ▶

2 Molly used base-ten blocks to model a division number sentence.

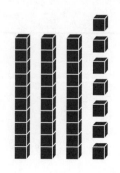 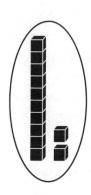 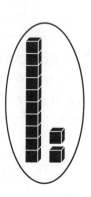 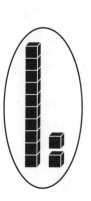

Which division sentence did she model?

(A) $32 \div 3 = 10 \text{ r2}$

(B) $36 \div 3 = 12$

(C) $37 \div 3 = 12 \text{ r1}$

(D) $38 \div 3 = 12 \text{ r2}$

3 Divide and check.

$7\overline{)98}$

(A) 12 r1

(B) 12 r5

(C) 13 r4

(D) 14

GO ON ▶

4 All 78 fourth-grade students from Winslow School went on a trip to the zoo. The zookeeper put students into groups of 3 to 5 students. All of the groups were the same size and no students were left over. How many students were in each group? How many groups were there?

On the lines below, explain how you can use the strategy *predict and test* to solve this problem.

_____

_____

_____

_____

_____

Use the strategy *predict and test* to solve the problem.

Show All Work

Answer _____ students in each group and _____ groups

5 The graph below shows the number of new books a city library just received.

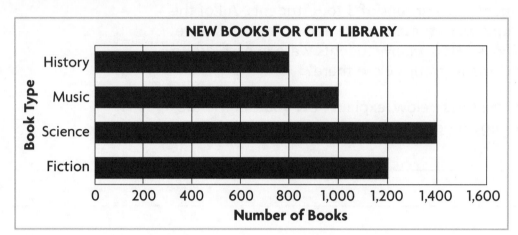

NEW BOOKS FOR CITY LIBRARY

The librarians will put the same number of science books in each of 7 bookcases. Which pattern can you use to find out how many science books will go in each bookcase?

Ⓐ 12 ÷ 6 = 2
 120 ÷ 6 = 20
 1,200 ÷ 6 = 200

Ⓑ 14 ÷ 7 = 2
 140 ÷ 7 = 20
 1,400 ÷ 7 = 200

Ⓒ 14 ÷ 2 = 7
 140 ÷ 2 = 70
 1,400 ÷ 2 = 700

Ⓓ 14 ÷ 1 = 14
 140 ÷ 1 = 140
 1,400 ÷ 1 = 1,400

GO ON ▶

6 Tad can write his full name in 5 seconds. He wrote the following equation to find out how many times he can write his name in 60 seconds.

$$60 \div 5 = n$$

Solve the equation for *n*.

(A) $n = 12$ times

(B) $n = 15$ times

(C) $n = 55$ times

(D) $n = 65$ times

7 There are 112 children signed up to take swimming lessons at the community pool. There will be 6 children in each group. ABOUT how many groups can be formed?

(A) about 10 groups

(B) about 15 groups

(C) about 20 groups

(D) about 25 groups

GO ON ▶

8 It took Greta a total of 109 minutes to do her homework for 5 different subjects. She worked on each subject about the same amount of time. ABOUT how long did she work on each subject?

(A) between 5 and 10 minutes each

(B) between 10 and 20 minutes each

(C) between 20 and 30 minutes each

(D) between 30 and 40 minutes each

9 Greg, Stan, and Jorge earned a total of $248 for working at the city park last month. They agreed to share their earnings equally. Greg estimated that they should each get about $100. What is a more REASONABLE estimate?

(A) about $60 each

(B) about $80 each

(C) about $90 each

(D) about $120 each

GO ON ▶

10 In which division problem are the digits in the quotient correctly placed with the correct answer?

(A) $\frac{21}{4\overline{)96}}$

(B) $\frac{24}{4\overline{)96}}$

(C) $\frac{21}{4\overline{)96}}$

(D) $\frac{24}{4\overline{)96}}$

11 During the 2001–2002 NBA season, Indiana and Toronto together won a total of 84 games. Both teams won the same number of games. How many basketball games did Indiana win during the 2001–2002 season?

(A) 21 games

(B) 32 games

(C) 40 games

(D) 42 games

GO ON ▶

12 Jenna bought flowers to create 4 table arrangements for a party. The pictograph shows how many of each type of flower she bought.

FLOWERS FOR TABLE ARRANGEMENTS

Tulips	✿ ✿ ✿
Daisies	✿ ✿
Tiger Lilies	✿ ✿ ✿ ✿ ✿
Daffodils	✿ ✿ ✿ ✿ ✿ ✿

Key: Each ✿ = 5 flowers.

Each arrangement has the same number of flowers. How many flowers are in each arrangement?

(A) 10 flowers   (C) 20 flowers

(B) 15 flowers   (D) 25 flowers

13 For their vacation, the Johnsons spent $720 to stay at a motel on the beach for 6 nights. How much did it cost for each night at the motel? What method would you use to solve?

(A) $110; method: pencil and paper

(B) $120; method: mental math

(C) $150; method: pencil and paper

(D) $160; method: mental math

© Harcourt

GO ON ▶

14 Andrea has a roll of ribbon to make bows for the bags of cookies she is selling at the school bake sale. The roll has 245 inches of ribbon. Each bow uses 8 inches of ribbon. Andrea used the following method to find out how many bows she can make.

$$8\overline{)245} ^{\underline{30\ r5}}$$

Andrea dropped the remainder and increased the quotient by 1. She thinks she can make 31 bows.

On the lines below, explain why Andrea's solution is incorrect.

_____

_____

_____

_____

_____

What should you do with the remainder in this problem?

Answer _____

How many bows can Andrea make?

Answer _____ bows

GO ON ▶

15 There are 524 students in Carver Elementary School. They are all going on a trip to Washington, D.C. The school has rented 11 buses for the trip. ABOUT how many students will travel on each bus?

(A) about 50 students

(B) about 60 students

(C) about 70 students

(D) about 80 students

16 Which number completes the pattern?

$$56 \div 8 = 7$$

$$560 \div 80 = 7$$

$$5,600 \div 80 = 70$$

$$56,000 \div 80 = \blacksquare$$

(A)   70

(B)   700

(C)   750

(D)   7,000

GO ON ▶

17 Mr. Morris taught his class how to divide using the model below:

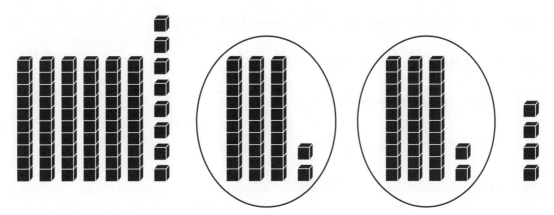

Which division sentence did Mr. Morris model?

(A) 32 ÷ 2 = 16

(B) 68 ÷ 2 = 34

(C) 68 ÷ 4 = 17

(D) 68 ÷ 32 = 2 r4

18 Which statement is true about the estimate for the first digit in the quotient?

$$\begin{array}{r} 9 \\ 46\overline{)375} \end{array}$$

(A) It is too high.

(B) It is too low.

(C) It is just right.

(D) It is placed above the wrong digit.

GO ON ▶

19 There are 92 magazines stacked in 23 equal piles.
How many magazines are in each pile?

Solve the problem and tell which operation you used.

(A)　　　4 magazines; division

(B)　　　69 magazines; subtraction

(C)　　115 magazines; addition

(D)　2,116 magazines; multiplication

20 Look at this number pattern.

96, 84, 72, ■, ■, 36

What is a rule for the pattern?

Answer _____

What are the missing numbers in the pattern?

Answer _____

STOP ■

Name _____

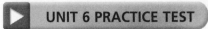
Getting Ready for the **ISTEP**+

1 Which geometric term describes this figure?

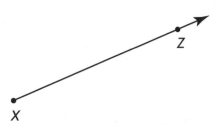

X

Z

A line segment *XZ*   **C** plane *XZ*

B ray *XZ*   **D** line *XZ*

2 Classify the angle of the window in the house.

A acute   **C** right

B obtuse   **D** straight

GO ON ▶

© Harcourt

3 What is the measure of angle *XYZ*?

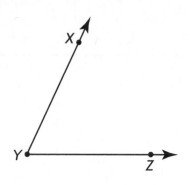

(A) ∠*XYZ* = 50°

(B) ∠*XYZ* = 65°

(C) ∠*XYZ* = 80°

(D) ∠*XYZ* = 85°

4 Indiana's first major railroad line connected Madison and Indianapolis. It was completed in 1847. Which term describes the railroad tracks?

(A) intersecting lines   (C) parallel lines

(B) obtuse angles   (D) perpendicular lines

GO ON ▶

5 Name two perpendicular lines in the figure below.

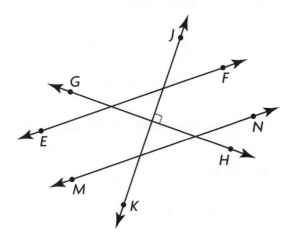

(A) line *GH* and line *JK*

(C) line *GH* and line *MN*

(B) line *EF* and line *MN*

(D) line *EF* and line *JK*

6 Sam walked 5 blocks east from his home to the library. Then he walked 3 blocks north to the video store. Then he walked 7 blocks west to the post office. Then he walked 4 blocks south and 2 blocks east to the farmer's market. What is the fastest way Sam can get home from the market?

(A) He can walk 1 block south.

(B) He can walk 2 blocks south.

(C) He can walk 1 block north.

(D) He can walk 2 blocks north.

GO ON ▶

7 Elena drew a quadrilateral with only one pair of parallel sides. Which figure did she draw?

(A) trapezoid

(C) rhombus

(B) parallelogram

(D) rectangle

8 These figures are all quadrilaterals.

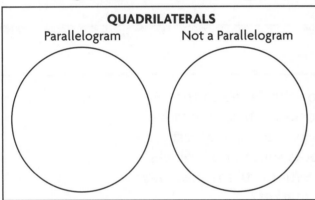

Sort the figures into the Venn diagram below.

QUADRILATERALS

Parallelogram   Not a Parallelogram

On the lines below, explain how you decided which figure to put in each circle.

_____

_____

_____

_____

_____

GO ON ▶

9 Which figure shows line symmetry?

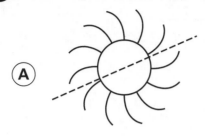

Ⓐ

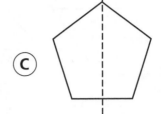

Ⓒ

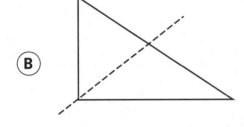

Ⓑ

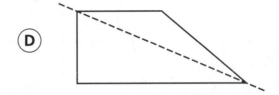

Ⓓ

10 This is a regular hexagon. Draw all its lines of symmetry.

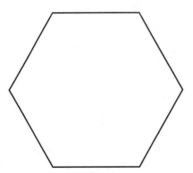

How many lines of symmetry are there in a regular hexagon?

Answer _____ lines of symmetry

GO ON ▶

11 Which shows a pair of CONGRUENT figures?

Ⓐ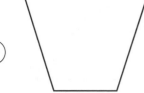

Ⓑ

Ⓒ

Ⓓ

12 How would you change Figure A to make it CONGRUENT to Figure B?

Figure A   Figure B

(A) Change the size of the angles.

(B) Change the length of the sides.

(C) Change the number of sides.

(D) Change the number of parallel sides.

GO ON ▶

13 Terry drew this logo for his family's hog farm in Indiana.

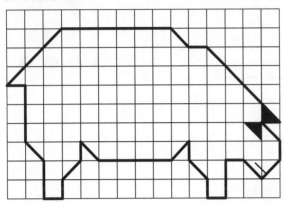

Terry wants to enlarge his logo to put it on a poster. Make a larger picture of his logo on the grid paper below.

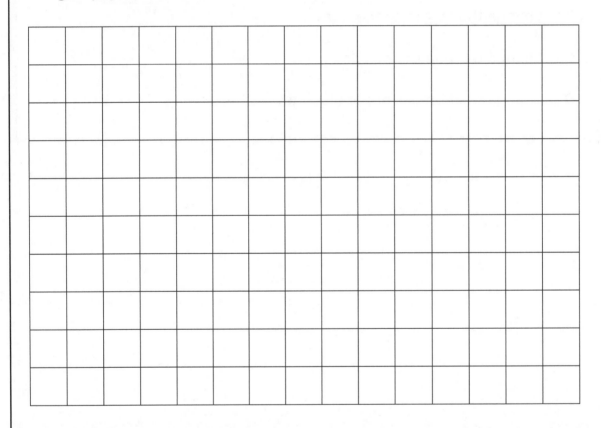

GO ON ▶

14 What is a rule for the pattern?

(A) Add 1 more dot to each figure.

(B) Take away 1 dot from each figure.

(C) Turn the figure upside down.

(D) Add 1 dot to the end of each row.

15 Which is probably the next figure in the pattern?

A ∀ ∀ ◁ A ?

(A) ∀

(C) A

(B) ◁

(D) ◁

16 Which number is represented by the letter *H* on the number line?

(A) ⁻7

(B) ⁻3

(C) 0

(D) ⁺7

17 Which number makes the inequality true?

$$g > 16$$

(A) 19

(B) 16

(C) 13

(D) 10

GO ON ▶

18 Which number makes the inequality true?

$k < 1$

Ⓐ   0

Ⓑ   4

Ⓒ   9

Ⓓ   13

19 Which number makes the inequality true?

$w \geq 5$

Ⓐ   1

Ⓑ   3

Ⓒ   4

Ⓓ   5

© Harcourt

GO ON ▶

20 Name three whole numbers that make the inequality true.

$m + 2 \geq 4$

Answer _____

_____

On the lines below, explain how you know if a number makes an inequality true.

_____

_____

_____

_____

On the number line below, graph the three whole numbers you named.

STOP ■

Getting Ready for the **ISTEP+**

1 Melissa colored flags in her Indiana state activity book.

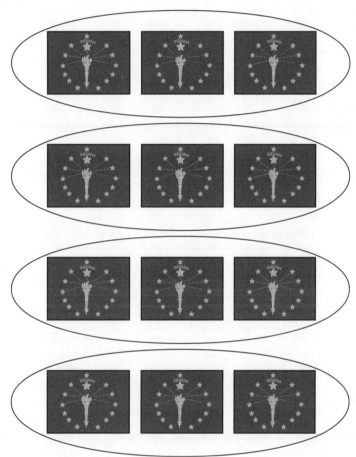

Which fraction names the part of the flags that she colored?

(A) $\frac{4}{3}$

(B) $\frac{4}{4}$

(C) $\frac{1}{3}$

(D) $\frac{1}{4}$

GO ON ▶

2 Six out of six kites are blue. Three out of five bikes are red. Which fraction shows the part of the kites that are blue?

(A) $\frac{3}{5}$

(B) $\frac{6}{6}$

(C) $\frac{6}{11}$

(D) $\frac{6}{9}$

3 Find the missing numerator.

$$\frac{\blacksquare}{4} = \frac{2}{2}$$

(A) 2

(B) 4

(C) 6

(D) 8

GO ON ▶

4 The table shows how far each of 3 students walks to school every day.

WALKING DISTANCES

Student	Distance Walked
Greg	$\frac{3}{4}$ mile
Jason	$\frac{3}{8}$ mile
Sara	$\frac{1}{2}$ mile

On the lines below, explain how you can use a number line to find out who walks the GREATEST distance to school and who walks the LEAST distance to school.

_____

_____

_____

_____

Solve the problem.

0   1

Answer greatest distance: _____

least distance: _____

© Harcourt

GO ON ▶

5 Clarence wrote a book report. The shaded parts of the model show the number of pages he wrote.

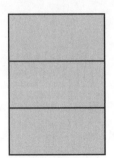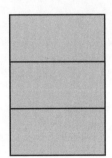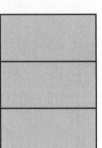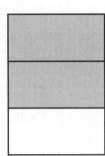

How long was Clarence's book report?

Ⓐ $\frac{4}{3}$ pages

Ⓑ $3\frac{1}{3}$ pages

Ⓒ $3\frac{2}{3}$ pages

Ⓓ $1\frac{2}{3}$ pages

6 Which number represents the number of dozens of eggs shown?

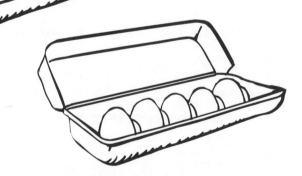

Ⓐ $1\frac{5}{12}$ dozen

Ⓑ $1\frac{7}{12}$ dozen

Ⓒ $\frac{5}{12}$ dozen

Ⓓ $\frac{7}{12}$ dozen

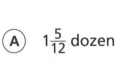

GO ON ▶

© Harcourt

7 Which fraction does the model show?

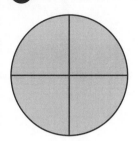 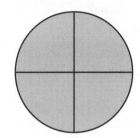 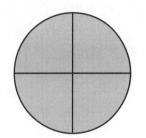 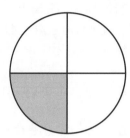

(A) $\frac{13}{4}$

(B) $\frac{4}{3}$

(C) $\frac{3}{4}$

(D) $\frac{4}{13}$

8 Sal feeds his dogs $2\frac{1}{5}$ cans of dog food each day. Which fraction shows the amount of food Sal feeds his dogs?

(A) $\frac{11}{2}$ cans

(B) $\frac{8}{3}$ cans

(C) $\frac{11}{5}$ cans

(D) $\frac{8}{5}$ cans

© Harcourt

GO ON ▶

9 There was 1 large can of juice for the class party. The boys drank $\frac{7}{10}$ of the juice. The girls drank $\frac{3}{10}$ of the juice. How much of the can of juice did the class drink at the party?

(A) $\frac{4}{10}$ of the can

(B) $\frac{10}{20}$ of the can

(C) $\frac{9}{10}$ of the can

(D) 1 can

10 Ramon drove a bus from South Bend to Richmond in $3\frac{1}{4}$ hours. He then drove from Richmond to Evansville in $4\frac{3}{4}$ hours. Which operation would you use to find the total time for the trip?

(A) addition

(B) subtraction

(C) multiplication

(D) division

11 Justin is $4\frac{1}{2}$ feet tall. He is $\frac{1}{2}$ foot taller than his younger sister, Alice. Which operation would you use to find Alice's height?

(A) addition

(B) subtraction

(C) multiplication

(D) division

GO ON ▶

12 Pablo rode his bike $\frac{1}{5}$ mile to the store. Then he rode another $\frac{3}{10}$ mile to his friend's house.

1

$\frac{1}{5}$	$\frac{1}{10}$	$\frac{1}{10}$	$\frac{1}{10}$

?

Use the fraction bars to find out how far Pablo rode.

A $\frac{4}{15}$ mile

C $\frac{1}{2}$ mile

B $\frac{4}{10}$ mile

D $\frac{7}{10}$ mile

13 Gia had math and spelling homework last night. She took $\frac{1}{2}$ hour to do her math homework and $\frac{1}{3}$ hour to do her spelling homework.

1

$\frac{1}{2}$	$\frac{1}{3}$

?

Use the fraction bars to find out how long it took Gia to do her homework.

A $\frac{2}{5}$ hour

C $\frac{2}{3}$ hour

B $\frac{2}{4}$ hour

D $\frac{5}{6}$ hour

© Harcourt

GO ON ▶

14 Gerald had $\frac{1}{2}$ gallon of milk. He used $\frac{3}{8}$ of the milk to make pancakes.

1

$\frac{1}{2}$

$\frac{1}{8}$	$\frac{1}{8}$	$\frac{1}{8}$	?

Use the fraction bars to find out how much milk is left.

(A) $\frac{1}{8}$ gallon   (C) $\frac{1}{4}$ gallon

(B) $\frac{1}{2}$ gallon   (D) $\frac{3}{8}$ gallon

15 Lyle is walking home from the library. He has walked $\frac{1}{4}$ mile. He lives $\frac{7}{8}$ mile from the library.

1

$\frac{1}{8}$	$\frac{1}{8}$	$\frac{1}{8}$	$\frac{1}{8}$	$\frac{1}{8}$	$\frac{1}{8}$	$\frac{1}{8}$

$\frac{1}{4}$	?

Use the fraction bars to find out how far Lyle still has to walk to get home.

(A) $\frac{1}{8}$ mile   (C) $\frac{5}{8}$ mile

(B) $\frac{3}{8}$ mile   (D) $\frac{3}{4}$ mile

GO ON ▶

© Harcourt

16 You are planning to do an experiment in which you toss a coin and at the same time toss a number cube labeled 1–6.

Make a table to show all the possible outcomes.

Coin	Number Cube					

How many possible outcomes are there?

Answer _____ possible outcomes

17 Hillary has a bag containing these shapes.

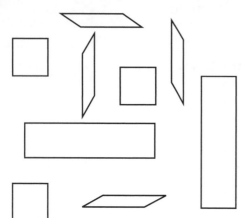

She chooses one shape from the bag without looking. What is the probability that the shape will be a quadrilateral?

Ⓐ impossible, $\frac{0}{9}$

Ⓒ likely, $\frac{7}{9}$

Ⓑ unlikely, $\frac{3}{9}$

Ⓓ certain, $\frac{9}{9}$

18 A spinner is divided into 8 equal sections labeled 1 to 8. The table shows the results of spinning the pointer 30 times. Find the fraction of spins when the pointer stopped on an even number.

Spinner Experiment (30 Spins)

Outcome	1	2	3	4	5	6	7	8
Tally	|||	||||	₩₩	||||	|||	||||	|||	||||

Ⓐ $\frac{8}{15}$

Ⓒ $\frac{7}{8}$

Ⓑ $\frac{7}{30}$

Ⓓ $\frac{1}{4}$

GO ON ▶

© Harcourt

19 Bill and his friends are planning a trip to the Indiana State Museum. They want to see a show at the IMAX Theater. They can go on Monday, Tuesday, or Friday. The show times are 10:00 A.M., 12:45 P.M., and 3:30 P.M.

Make a diagram to show the different combinations of day and time.

How many different combinations are possible?

Answer _____ combinations

GO ON ▶

20 Hector and Anna are playing a game with a number cube labeled 1 to 6. It is Hector's turn. To win the game he must roll two numbers that have a sum of 6.

Make an organized list to show all the possible outcomes when Hector rolls the number cube twice.

On your list above, circle all the outcomes that would allow Hector to win the game.

How many different ways can Hector win?

Answer _____ ways

STOP ■

Getting Ready for the **ISTEP+**

1 The northern cardinal is the Indiana state bird. Measure a tail feather to the nearest $\frac{1}{4}$ inch.

(A) $3\frac{1}{4}$ inches

(B) $3\frac{1}{2}$ inches

(C) $3\frac{3}{4}$ inches

(D) 4 inches

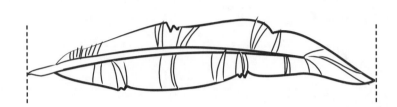

2 Soybeans are a major part of Indiana agriculture. Soybean oil, soy flour, and soy protein are used in many kinds of food. Soybeans are also used as animal feed. Measure the soybean pod to the nearest $\frac{1}{8}$ inch.

(A) 2 inches

(C) $2\frac{1}{4}$ inches

(B) $2\frac{1}{8}$ inches

(D) $2\frac{3}{8}$ inches

© Harcourt

GO ON ▶

3 Which equation can be used to complete the table?

Yards, y	2	4	6	8	10	12
Feet, f	6	12	18	■	■	■

(A) $f = y \times 3$

(B) $y = f \times 3$

(C) $f = y + 4$

(D) $y = f \div 2$

4 Jon covers a shelf with a sheet of plastic. The shelf is 70 inches long. The roll of plastic is 6 feet long. The plastic is the same width as the shelf. How many inches of plastic does Jon have left?

(A) 1 inch

(B) 2 inches

(C) 3 inches

(D) 4 inches

© Harcourt

GO ON ▶

5 Change the unit. Which is the missing number in the table?

Quart	4	8	12	16	20	⬛
Gallon	1	2	3	4	5	6

A 22

B 24

C 25

D 28

6 James and his dad are making pancakes for 25 campers. They need 1 cup of buttermilk for each batch of pancakes. How many batches can they make from 2 quarts of buttermilk?

Make a table to solve the problem.

Answer _____ batches

GO ON ▶

7 In 2002 Indiana farmers produced more than 250 million pounds of popcorn. Use a centimeter ruler to measure the width of the popcorn kernel.

(A) 12 millimeters

(B) 12 centimeters

(C) 12 decimeters

(D) 12 meters

8 A bulletin board is 3 meters wide. Tammy will use half of it for a display about Indiana. How wide will her display be in centimeters?

(A) 54 centimeters

(B) 100 centimeters

(C) 125 centimeters

(D) 150 centimeters

GO ON ▶

9 Karen and Marty each have some juice to bring to a party. The two have a total of 11 liters of juice. Marty has 2 liters more than twice the amount of juice that Karen has. How much juice do they each have?

Draw a diagram to solve.

Write your answer on the lines below.

Answer Marty: _____ liters

Karen: _____ liters

GO ON ▶

10 Which decimal shows the same amount as $\frac{70}{100}$?

(A) 0.07

(B) 0.17

(C) 0.70

(D) 0.77

11 Which fraction shows the same amount as 0.25?

(A) $\frac{1}{4}$

(B) $\frac{2}{5}$

(C) $\frac{1}{2}$

(D) $\frac{25}{10}$

GO ON ▶

12 Harriet feeds corn to her chickens. The decimal model shows how many sacks of corn she has left.

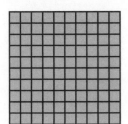 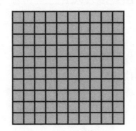 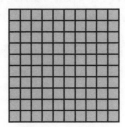 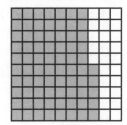

Which mixed number and decimal represent the shaded portion of this model?

(A) $3\frac{1}{2}$, 3.12

(B) $3\frac{1}{2}$, 3.50

(C) $3\frac{3}{4}$, 3.34

(D) $3\frac{3}{4}$, 3.75

GO ON ▶

13 Dominic, Samson, and Troy are entered in a race.
Their times were 15.10 sec, 14.65 sec, and 11.90 sec.
Troy was the fastest runner. Samson ran slower
than Dominic. How fast did each boy run?

Use logical reasoning. Organize what you
know in a table. Show all the possibilities.

Write your answers below.

Answer Dominic: _____ sec

Samson: _____ sec

Troy: _____ sec

GO ON ▶

14 Selena finished a race in 18.26 seconds. Round her time to the nearest tenth of a second.

(A) 19.0 seconds

(B) 18.3 seconds

(C) 18.2 seconds

(D) 18.0 seconds

15 Use the model to help find the sum.

0.35 + 0.81

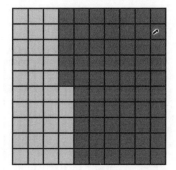 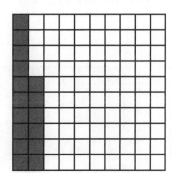

(A) 1.16

(B) 1.18

(C) 1.35

(D) 1.81

GO ON ▶

16 Find the sum.

$$\begin{array}{r} 3.89 \\ + 2.46 \\ \hline \end{array}$$

(A) 5.25

(B) 5.35

(C) 6.25

(D) 6.35

17 Use the model to help find the difference.

2.4 − 1.9

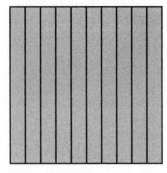 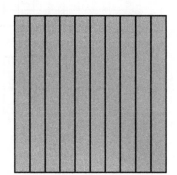 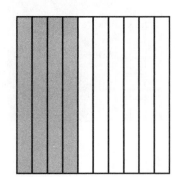

(A) 0.5

(B) 1.4

(C) 1.5

(D) 2.9

GO ON ▶

Name _____

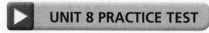

18 Find the difference.

$$\begin{array}{r} 6.47 \\ -\ 1.88 \\ \hline \end{array}$$

(A) 4.58

(B) 4.59

(C) 5.48

(D) 5.49

19 Find the difference.

$$\begin{array}{r} \$20 \\ -\ \$17.99 \\ \hline \end{array}$$

(A) $2.01

(B) $2.09

(C) $3.01

(D) $3.09

© Harcourt

GO ON ▶

20 At the China Café, Ling ordered egg drop soup, vegetable lo mein, chicken wings, and juice. How much was her bill?

China Café

Egg Drop Soup	$2.35
Pork Stir-fry	$6.99
Beef Stir-fry	$7.49
Vegetable Lo-mein	$5.75
Chicken Wings	$4.79
Juice	$1.25
Tea	$0.95

ESTIMATE the answer by rounding to the nearest dollar.

Answer $ _____

Now, solve the problem.

Show All Work

Answer $ _____

Is your answer reasonable? On the lines below, explain how you know.

_____

_____

© Harcourt

STOP ■

Getting Ready for the **ISTEP+**

1 Which formula could you use to find the perimeter of the polygon shown? Let *P* stand for perimeter.

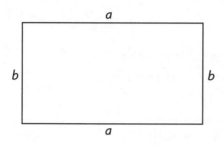

(**A**) $P = (2 \times a) + (2 \times b)$

(**B**) $P = a \times b$

(**C**) $P = a + b$

(**D**) $P = 4 \times a$

2 The baseball field for Little League is a square. Each side is 60 feet long. For Senior League, each side of the square field is 90 feet long. How much greater is the perimeter of a Senior League field than that of a Little League field?

Perimeter of a square = 4 × side

(**A**) 150 feet   (**C**) 90 feet

(**B**) 120 feet   (**D**) 60 feet

GO ON ▶

3 Lane's Swimming Pools builds rectangular pools in different sizes.

RECTANGULAR POOLS	
Length (in feet)	Width (in feet)
28	15
32	16
36	18

What is the difference in the perimeters of the pool with the GREATEST perimeter and the pool with the LEAST perimeter?

Perimeter = (2 × length) + (2 × width)

(A) 22 feet

(B) 32 feet

(C) 86 feet

(D) 108 feet

4 One side of a square bedroom is 9 feet long. What length of wallpaper is needed to make a border all around the room?

(A) 18 feet

(B) 21 feet

(C) 27 feet

(D) 36 feet

© Harcourt

GO ON ▶

5 The total distance from Indianapolis to Philadelphia, from Philadelphia to Memphis, and from Memphis back to Indianapolis is 2,068 miles.

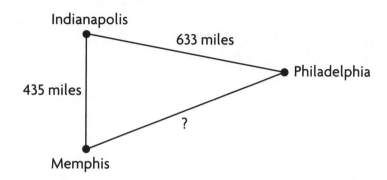

On the lines below, explain how to find the distance from Philadelphia to Memphis.

_____

_____

_____

_____

_____

Solve the problem. Write your answer on the line below.

Show All Work

_____

_____

_____

Answer _____ miles

GO ON ▶

6 A rectangular garden is 24 feet long and 12 feet wide. A gardener is putting an iron border around the garden. Each section of the border is 3 feet long and costs $26.

On the lines below, explain how you could use a formula to help find the cost of the border.

_____

_____

_____

_____

_____

Use your method to solve the problem.

Show All Work

Answer $ _____

If the gardener extends the length of the garden another 6 feet, how much more will it cost to put a border around the larger garden?

Answer $ _____

GO ON ▶

7 Look at the two tablecloths below.

70 inches

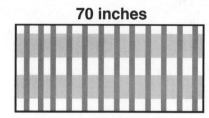

54 inches

72 inches

90 inches

How much more area can be covered with the larger tablecloth than with the smaller tablecloth?

Area = length × width

(**A**) 2,700 square inches   (**C**) 76 square inches

(**B**) 1,350 square inches   (**D**) 38 square inches

8 How much carpet do you need to completely cover the floor of a square den that is 11 feet long?

(**A**) 44 square feet

(**B**) 110 square feet

(**C**) 121 square feet

(**D**) 621 square feet

GO ON ▶

9 Vera measures the tennis court at the school. She says it is 36 wide and 78 long, but she left out the unit of measure she used. Which is a reasonable estimate for the area of the tennis court?

> Area = length × width

(A) about 3,000 square miles

(B) about 3,000 square inches

(C) about 3,000 square centimeters

(D) about 3,000 square feet

10 Find the area of the irregular polygon.

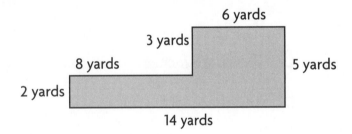

(A) 24 square yards

(B) 38 square yards

(C) 46 square yards

(D) 84 square yards

GO ON ▶

11 Mitch wants to carpet a living room, dining room, and hall with the same wall-to-wall carpeting.

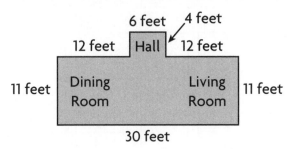

6 feet
4 feet
12 feet | Hall | 12 feet
11 feet | Dining Room | Living Room | 11 feet
30 feet

Use the diagram. Find out how much carpeting he will need.

Area = length × width

(A) 90 square feet

(C) 330 square feet

(B) 132 square feet

(D) 354 square feet

12 Which statement is true about the figures below?

8 meters

4 meters

16 meters

2 meters

Perimeter = (2 × length) + (2 × width)
Area = length × width

(A) They have the same perimeters but different areas.

(B) Their perimeters and areas are different.

(C) They have the same areas but different perimeters.

(D) They have the same areas and perimeters.

© Harcourt

GO ON ▶

13 Jana drew this square.

12 inches

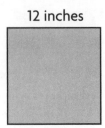

Which figure has the same perimeter but a different area?

| Perimeter = (2 × length) + (2 × width)
Area = length × width |

(A)   12 inches
2 inches

(B)   16 inches
8 inches

(C)   6 inches

(D)   16 inches
9 inches

GO ON ▶

14 Mr. Chung is laying square tiles on the floor of his kitchen and screened porch. The kitchen is 10 feet long and 12 feet wide. The screened porch is 8 feet long and 15 feet wide. Which statement is true?

Area = length × width

(A) He will need the same number of tiles for the kitchen as for the screened porch.

(B) He will need more tiles for the kitchen than for the screened porch.

(C) He will need more tiles for the screened porch than for the kitchen.

(D) He will need twice as many tiles for the screened porch as for the kitchen.

15 The table below shows the different-size area rugs sold by Carpet Outlet.

CARPET OUTLET AREA RUGS			
Rug	Length (in feet)	Width (in feet)	Area (in square feet)
Small	4	2	8
Medium	8	4	32
Large	16	8	128

What pattern do you see in the lengths and widths of the rugs they sell?

Answer _____

What pattern do you see in the areas of the rugs?

Answer _____

GO ON ▶

16 Sandy constructed a triangular pyramid out of cardboard. Which shapes did she use?

(A) 3 triangles and 1 rectangle

(B) 4 triangles

(C) 5 triangles

(D) 3 triangles and 1 square

17 Which solid figure can you make with this net?

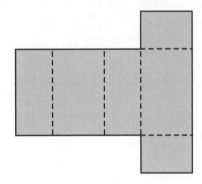

(A) rectangular pyramid

(B) rectangular prism

(C) triangular pyramid

(D) triangular prism

© Harcourt

GO ON ▶

18 What is the volume of the prism below?

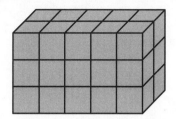

Volume = length × width × height

(A) 30 cubic units   (C) 16 cubic units

(B) 25 cubic units   (D) 15 cubic units

19 A tank with a volume of 1,000 cubic centimeters can hold 1 liter of water. What is the capacity of the fish tank shown below?

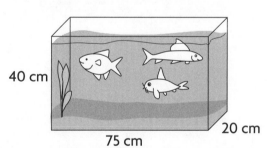

40 cm

75 cm

20 cm

Volume = length × width × height

(A) 6 liters   (C) 60 liters

(B) 30 liters   (D) 135 liters

GO ON ▶

20 Mr. Rothman packs 10 books in a shipping box. The box is 6 inches high, 18 inches long, and 12 inches wide. It costs $25.80 to ship the box of books. What is the volume of the shipping box?

What information do you need to solve the problem? Is there too much or too little information? Explain.

_____

_____

_____

_____

Now, solve the problem.

Show Your Work

Answer _____ cubic inches

STOP ■

VOCABULARY REVIEW

Number Sense Vocabulary

Write the word or words that best complete
each sentence. Some words will not be used.

decimal	fraction	greater than	hundreds
millions	mixed number	less than	hundredth
period	place-value	thousands	tenth

1. The _____ equivalent for $\frac{1}{2}$ is 0.5.

2. The digit in the _____ place in the number 12,345,678 is 2.

3. The three-digit group of 512 is in the thousands _____ in the number 6,512,904.

4. You can rewrite the whole number 5 as the _____ $\frac{15}{3}$.

5. The number 8.60 is _____ the number 8.5.

6. $4\frac{2}{3}$ is called a _____.

7. The number 615 is _____ the number 621.

8. The digit in the _____ place in the number 3,642,781 is 7.

9. When 8.15 is rounded to the nearest _____, it becomes 8.2.

10. The _____ position of the digit 6 in the number 162,195 is ten thousands.

Computation Vocabulary

For each term in Column 1, find the term in Column 2 that is most nearly the *opposite* in meaning. Write the letter of the opposite term in the space provided.

Column 1	Column 2
1. division _____	A numerator
2. unlike fractions _____	B quotient
3. product _____	C multiplication
4. denominator _____	D like fractions
5. exact answer _____	E estimate

Write the letter of the word or words that best match the example.

6. $\frac{1}{2}$ _____

7. $4 \times 0 = 0$ _____

8. $12 \times 1 = 12$ _____

9. 0.5 _____

10. $25 + 0 = 25$ _____

A   **Identity Property of Multiplication**

B   **Identity Property of Addition**

C   **Zero Property of Multiplication**

D   **fraction**

E   **decimal**

Name _____

Algebra and Functions Vocabulary

Choose the correct word or words in parentheses to complete each sentence.

1. The first operation to perform in the expression $4 + 10 \div 2$ is _____ (addition, division).

2. In $3 + n = 7$, n is a(n) _____ (operation, variable).

3. In the expression $9 \div 3 \times 2$, first find the _____ (product, quotient).

4. Multiplication and division are _____ (inverse operations, order of operations).

5. An example of an _____ (expression, equation) is $20 + n$.

6. $21 > 15 + x$ is an _____ (inverse operation, inequality).

7. Marci receives an allowance of $2 per week. To find the total amount she will receive in 15 weeks, she should use the operation _____ (multiplication, division).

8. $15 - y = 12$ is an example of an _____ (expression, equation).

9. The correct _____ (order of operations, inverse operations) to use to find $12 - 6 \div 2$ is to divide and then subtract.

10. The symbols used to show which operations in an expression should be performed first are _____ (inequalities, parentheses).

Geometry Vocabulary

Write the letter of the figure that best matches
each word or words. Use each letter once.

1. **rhombus**

   _____

2. **acute angle**

   _____

3. **perpendicular lines**

   _____

4. **rectangle**

   _____

5. **ray**

   _____

6. **right angle**

   _____

7. **parallel lines**

   _____

8. **square**

   _____

9. **vertex**

   _____

10. **trapezoid**

    _____

A

B

C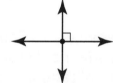

D

E

F

G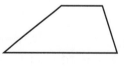

H

I

J

Measurement Vocabulary

Tell whether each statement is *true* or *false*. If it is false, change the underlined word to make the statement true.

1. There are 12 inches in 1 <u>yard</u>.

   _____

2. The <u>perimeter</u> is the distance around a figure.

   _____

3. The <u>area</u> of a square with 5-inch sides is 25 square inches.

   _____

4. A gallon is a measure of <u>weight</u>.

   _____

5. One centimeter is <u>longer</u> than 1 millimeter.

   _____

6. There are 60 seconds in 1 <u>hour</u>.

   _____

7. The <u>volume</u> of a baseball bat is about 3 feet.

   _____

8. There are 1,000 meters in 1 <u>kilometer</u>.

   _____

9. A liter is a measure of <u>capacity</u>.

   _____

10. Multiply the length, width, and height to find the <u>area</u> of a box.

    _____

Data Analysis and Probability Vocabulary
Write the letter of the word or words that best complete each sentence.

1. You can show how data change over time by using a(n) _____.

2. A graph used to compare similar kinds of data is a(n) _____.

3. The distance between two numbers on the scale of a graph is the _____.

4. On a graph, a(n) _____ can be found where data increase, decrease, or stay the same over time.

5. A(n) _____ is a series of numbers placed at fixed distances on a graph to help label a graph.

6. A graph that uses pictures to show and compare information is a(n) _____.

7. A(n) _____ is a data display that shows groups of data arranged by place value.

8. Graphs are used to display _____, or information collected about people or things.

9. A(n) _____ can show data as a whole made up of different parts.

10. A(n) _____ can be used to gather information to record data.

A data

B pictograph

C survey

D interval

E scale

F stem-and-leaf plot

G double-bar graph

H line graph

I trend

J circle graph

Getting Ready for the **ISTEP+**

1 On the chalkboard, Sam wrote the expanded form of the number of square miles that Lake Michigan covers. What is the standard form of this number?

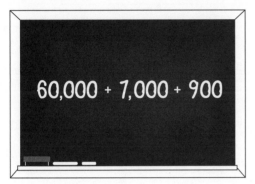

60,000 + 7,000 + 900

(A) 679,000

(B) 67,900

(C) 6,790

(D) 679

GO ON ▶

2 About 13,770 people live in Bedford; about 17,450 people live in Huntington; about 16,660 people live in Frankfort; and about 12,070 people live in Auburn.

Make a table. Organize the information from GREATEST to LEAST population.

Which of these cities in Indiana has the GREATEST population?

Answer _____

Order the populations using the less than symbol.

Answer _____

GO ON ▶

3 Which number rounds to 25,000 when rounded to the nearest thousand?

Ⓐ 24,095

Ⓑ 24,495

Ⓒ 25,495

Ⓓ 25,950

4 Find the difference.

4,300
− 2,479

Ⓐ 1,221

Ⓑ 1,821

Ⓒ 2,129

Ⓓ 2,131

GO ON ▶

5 Look at the input/output table below.

INPUT	n	5	10	15	20	25	30
OUTPUT	m	16	21	26	31	■	■

Write a rule for the input and output values.

Answer _____

Write the rule as an equation.

Answer _____

Use your equation to extend the pattern and complete the table. Write the missing output values below.

Answer _____

GO ON ▶

6 The students in Mr. Kendall's class read and record the temperature each day at noon. The line graph shows the changes in the temperatures over five days.

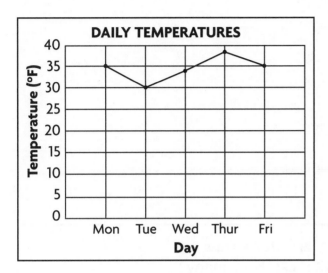

On which day was the temperature the HIGHEST?

(A) Monday

(B) Tuesday

(C) Wednesday

(D) Thursday

7 Find the value of the variable.

$s \div 8 = 8$

(A) $s = 1$

(B) $s = 8$

(C) $s = 16$

(D) $s = 64$

8 Mrs. Windsor's class orders 4 pizzas for a class party. Each pizza is cut into 12 slices. How many slices of pizza are there in all?

On the lines below, explain which operation you could use to solve this problem and why.

_____

_____

Write an equation you could use to solve the problem.

Answer _____

Use your equation to solve the problem.

Answer _____ slices

© Harcourt

GO ON ▶

9 Use the order of operations to find the value of the expression.

$$10 \times 5 - 4$$

(A) 10

(B) 20

(C) 46

(D) 64

10 A farmer picks 120 ears of corn. He puts the same number of ears of corn into each of 10 bags. Which equation would you use to find out how many ears of corn are in each bag?

(A) $120 \div 10 = e$

(B) $120 \times 10 = e$

(C) $120 \times e = 10$

(D) $10 + e = 120$

GO ON ▶

11 Look at the pattern below. What number is missing from the pattern?

$9 \times 3 = 27$

$9 \times 30 = 270$

$9 \times 300 = 2{,}700$

$9 \times 3{,}000 = \blacksquare$

(A) 270,000

(B) 27,000

(C) 20,700

(D) 2,700

12 Find the product.

$$\begin{array}{r} 47 \\ \times\, 5 \\ \hline \end{array}$$

(A) 125

(B) 202

(C) 205

(D) 235

© Harcourt

GO ON ▶

13 Erik went to the Parke County Covered Bridge Festival and bought photographs of all 32 covered bridges. He made a scrapbook with 7 photographs of each bridge. How many photographs does Erik have in his scrapbook?

Write an equation that you can use to solve the problem.

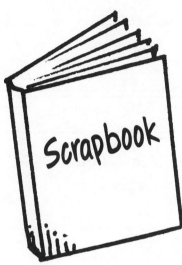

Answer _____

Now solve your equation. Write the answer below.

Answer _____ photographs

© Harcourt

14 There are 348 students at Lincoln Elementary School who eat lunch in the cafeteria. Lunch is served 20 days each month. How many lunches are served each month at Lincoln Elementary School?

On the lines below, explain how you could find the answer using mental math.

_____

_____

_____

_____

Show how you would solve this problem. Write your answer on the line below.

Show All Work

Answer _____ lunches

GO ON ▶

15 Jon, Josh, and Jordan divided 47 tickets for rides at the fair among themselves. Which model shows the division?

A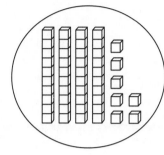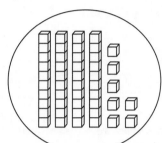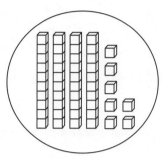

B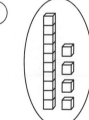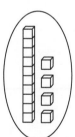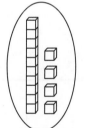

C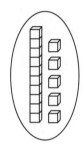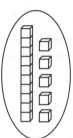

D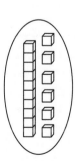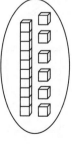

GO ON ▶

16 Divide.

$4\overline{)76}$

(A) 19

(C) 17

(B) 18

(D) 16

17 Which pair of figures appear to be CONGRUENT?

(A)

(B)

(C)

(D)

GO ON ▶

© Harcourt

18 Ben cut out 4 congruent circles. Then he cut each circle into 4 equal pieces. He used 3 of the pieces in a design for his notebook. He has $3\frac{1}{4}$ circles left. Which fraction shows the number of circles that are left?

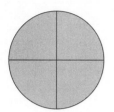 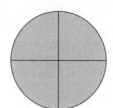 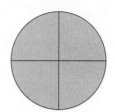 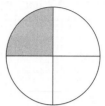

(A) $\frac{4}{4}$ (C) $\frac{13}{4}$

(B) $\frac{12}{4}$ (D) $\frac{16}{4}$

19 Use the fraction bars to add.

$\frac{1}{2} + \frac{3}{5}$

1			

| $\frac{1}{2}$ | | $\frac{1}{5}$ | $\frac{1}{5}$ | $\frac{1}{5}$ |

| $\frac{1}{10}$ | $\frac{1}{10}$ | $\frac{1}{10}$ | $\frac{1}{10}$ | $\frac{1}{10}$ | $\frac{1}{10}$ | $\frac{1}{10}$ | $\frac{1}{10}$ | $\frac{1}{10}$ | $\frac{1}{10}$ | $\frac{1}{10}$ |

$\frac{4}{5}$ 1 $\frac{11}{10}$ $\frac{12}{10}$

(A)   (B)   (C)   (D)

GO ON ►

20 Children must be 4 feet tall to ride the roller coaster. Maria is 46 inches tall. How many inches does she have to grow before she can ride the roller coaster?

| 1 foot = 12 inches |

(A) 1 inch

(B) 2 inches

(C) 3 inches

(D) 4 inches

21 The swing at the playground is $\frac{3}{4}$ meter from the ground. Which decimal shows how far the swing is from the ground?

(A) 0.25 meter

(B) 0.34 meter

(C) 0.75 meter

(D) 0.95 meter

GO ON ▶

22 Find the sum.

$$
\begin{array}{r}
53.56 \\
+ \ 8.9 \\
\hline
\end{array}
$$

(A) 51.46

(B) 52.46

(C) 61.46

(D) 62.46

23 The original Statehouse was built of Indiana limestone in Corydon. It was a square building with a perimeter of 160 feet. What was the length of each side?

| Perimeter = 4 × side |

(A) 40 feet

(B) 45 feet

(C) 60 feet

(D) 80 feet

GO ON ▶

24 The drawing shows the sidewalk in front of a movie theater. What is the area of the sidewalk?

$$\boxed{\text{Area} = \text{length} \times \text{width}}$$

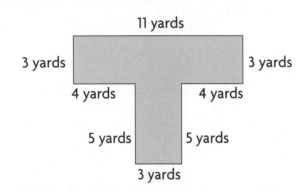

11 yards

3 yards   3 yards

4 yards   4 yards

5 yards   5 yards

3 yards

A 36 square yards

C 60 square yards

B 48 square yards

D 88 square yards

25 Find the volume of the rectangular prism.

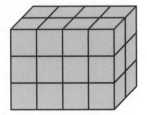

$$\boxed{\text{Volume} = \text{length} \times \text{width} \times \text{height}}$$

A 24 cubic units

C 16 cubic units

B 20 cubic units

D 12 cubic units

STOP ■